ELECTRIC ANIMALS

ELECTRIC EELS MAKE ELECTRICITY!

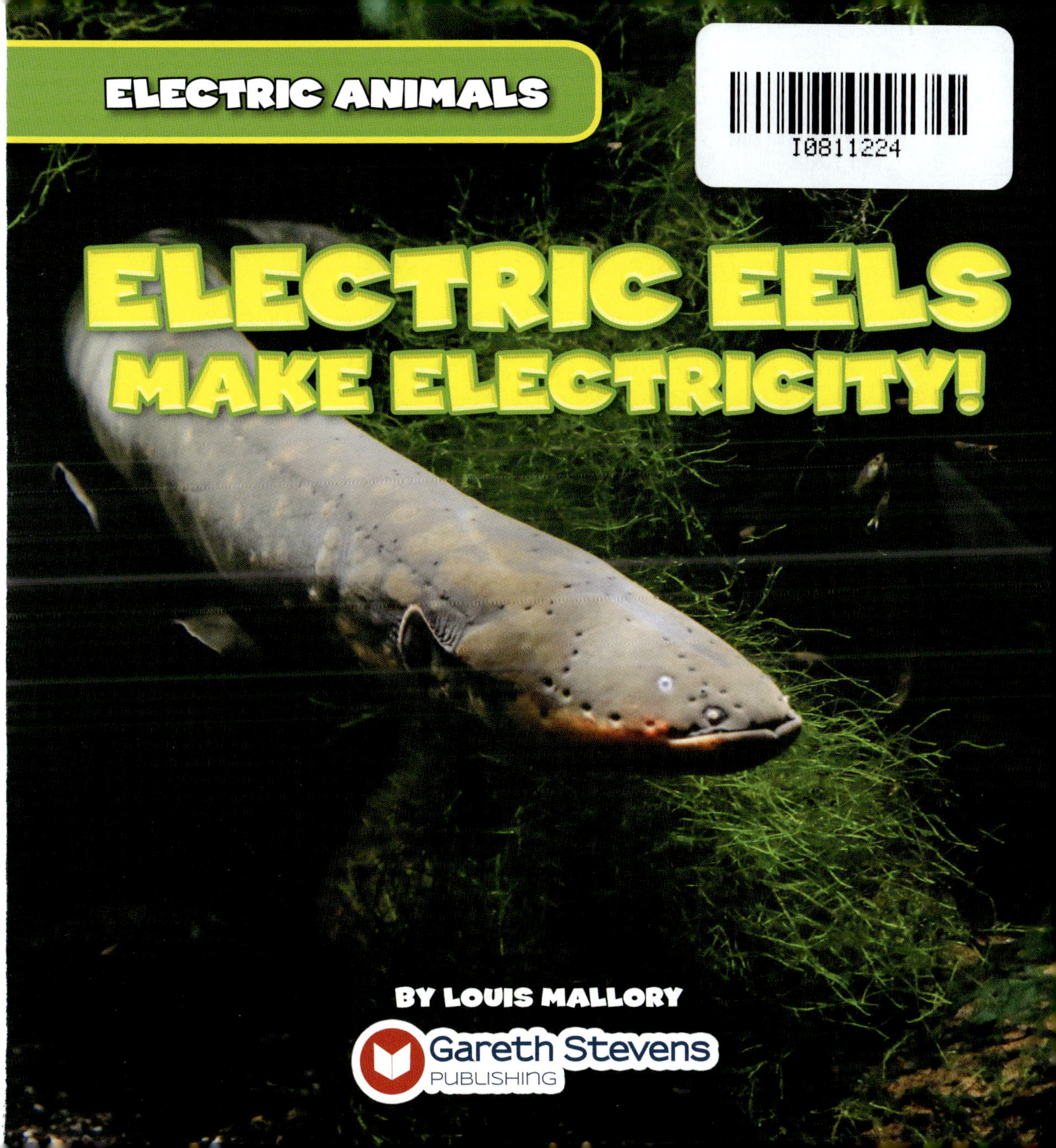

BY LOUIS MALLORY

Gareth Stevens PUBLISHING

Please visit our website, www.garethstevens.com. For a free color catalog of all our high-quality books, call toll free 1-800-542-2595 or fax 1-877-542-2596.

Cataloging-in-Publication Data
Names: Mallory, Louis.
Title: Electric eels make electricity! / Louis Mallory.
Description: New York : Gareth Stevens Publishing, 2024. | Series: Electric animals | Includes glossary and index.
Identifiers: ISBN 9781538292938 (pbk.) | ISBN 9781538292945 (library bound) | ISBN 9781538292952 (ebook)
Subjects: LCSH: Electric eel–Juvenile literature. | Electrophoridae–Juvenile literature.
Classification: LCC QL638.E34 M355 2024 | DDC 597'.48–dc23

Published in 2024 by
Gareth Stevens Publishing
2544 Clinton Street
Buffalo, NY 14224

Designer: Claire Wrazin
Editor: Natalie Humphrey

Photo credits: Cover, p. 1 Kseniia Mnasina/Shutterstock.com; background (series art) Romashka2/Shutterstock.com; pp. 5, 11, 21 tristan tan/Shutterstock.com; p. 7 VectorMine/ Shutterstock.com; p. 9 Language Of Dreams/Shutterstock.com; p. 13 clayton harrison/ Shutterstock.com; pp. 15, 17 (main), 19 Vladimir Wrangel/Shutterstock.com; p. 17 (inset) Galina Savina/Shutterstock.com.

Printed in the United States of America

CPSIA compliance information: Batch #CW24GS: For further information contact Gareth Stevens, New York, New York at 1-800-542-2595.

CONTENTS

Boldface words appear in the glossary.

The Electric Eel

You can find the electric eel in the Orinoco and Amazon Rivers in South America. This predator hunts for food in muddy, slow-moving water. But how does it catch its dinner? The electric eel shocks its food with **electricity**!

A Shocking Body

An electric eel's heart, **lungs**, and stomach are in and around its head. The rest of an eel's long body is made of three **organs**: the Sach's organ, main organ, and Hunter's organ. All three of these organs make electricity!

ELECTRIC ORGANS

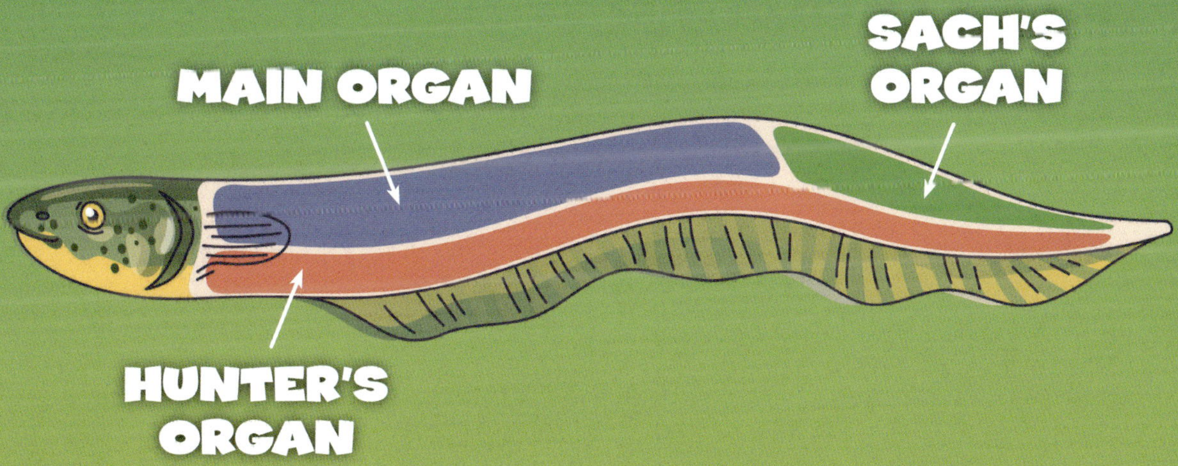

The Sach's organ, main organ, and Hunter's organ are made of special cells. These cells are called electrocytes. The electrocytes make electricity. One electric eel can have thousands of electrocytes! The electrocytes work together to make a strong electrical charge.

How Many Volts?

Scientists measure electrical energy in volts. Most electric eels can make over 600 volts of electricity. That's enough of a charge to stop a human heart! Some electric eels can be even stronger!

Hunting with Electricity

Electric eels use electricity to hunt for their dinner. Because electric eels can't see very well, they use small electrical **pulses** to find their prey, or the animals they hunt. These pulses, made by the Sach's organ, create an **electrical field** around the electric eel.

When prey crosses an electric eel's electrical field, the electric eel **attacks**! Using its Hunter's and main organs, it can make a powerful charge. This charge stuns the eel's prey, or makes the prey stop moving. Then the electric eel swallows it whole!

Dinner for an Eel

Electric eels are carnivores. This means an electric eel's favorite food is meat. They usually eat fish, bugs, frogs, lizards, and small **mammals** that get too close. Young electric eels will even eat the eggs of other electric eels!

Staying Safe

Electric eels don't just use their electricity to hunt. Electric eels also use their electricity to **protect** themselves. If a hungry predator gets too close, the electric eel will give them a shock!

Finding Other Eels

Electric eels use small pulses to **communicate** with other electric eels. These pulses share everything about the electric eel with the other eels. But electric eels need to be careful. A charge too strong can shock the other eel!

GLOSSARY

attack: To try to harm someone or something.

communicate: To share thoughts or feelings by sound, movement, or writing.

electrical field: An area of force surrounding a body with electrical charge.

electricity: A kind of energy that flows and is made by the movements of animals.

lungs: A part of the body used by animals to breathe.

mammal: A warm-blooded animal that has a backbone and hair, breathes air, and feeds milk to its young.

organ: A part inside an animal's body.

pulse: To make regular beats.

protect: To keep safe.

FOR MORE INFORMATION

BOOKS

Humphrey, Natalie. *Electric Eels Are Strange.* New York, NY: Enslow Publishing, 2023.

Wilson, Sierra. *Electric Eels.* New York, NY: AV2, 2022.

WEBSITES

National Aquarium: Electric Eel
www.aqua.org/explore/animals/electric-eel
Learn more about how electric eels live in the wild.

National Geographic Kids: Electric Eel
www.kids.nationalgeographic.com/animals/fish/facts/electric-eel
Discover more about where the electric eel lives, how it uses its electricity, and more!

Publisher's note to educators and parents: Our editors have carefully reviewed these websites to ensure that they are suitable for students. Many websites change frequently, however, and we cannot guarantee that a site's future contents will continue to meet our high standards of quality and educational value. Be advised that students should be closely supervised whenever they access the internet.

INDEX

leveled reader science

ELECTRIC ANIMALS

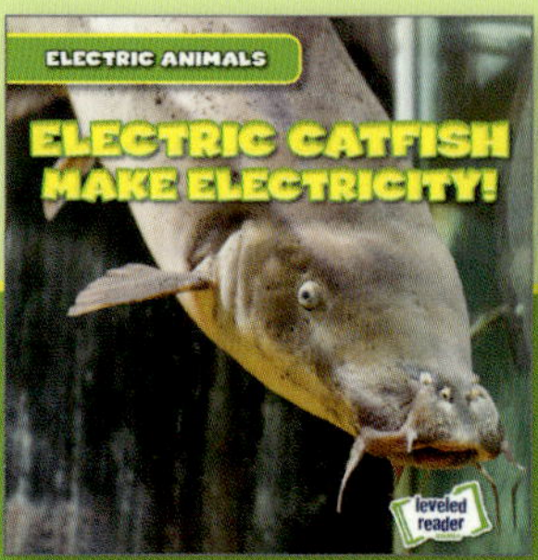

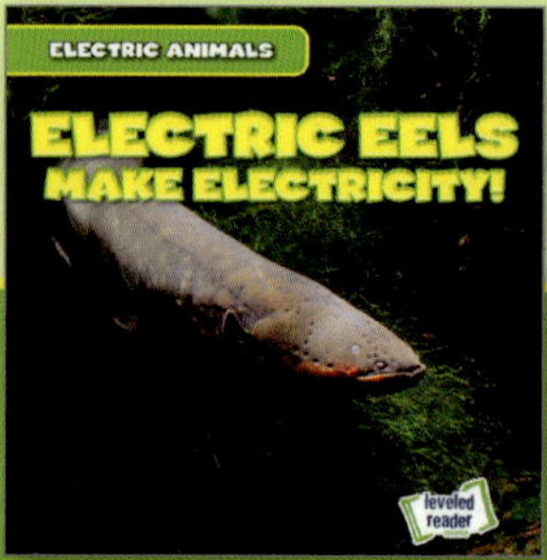

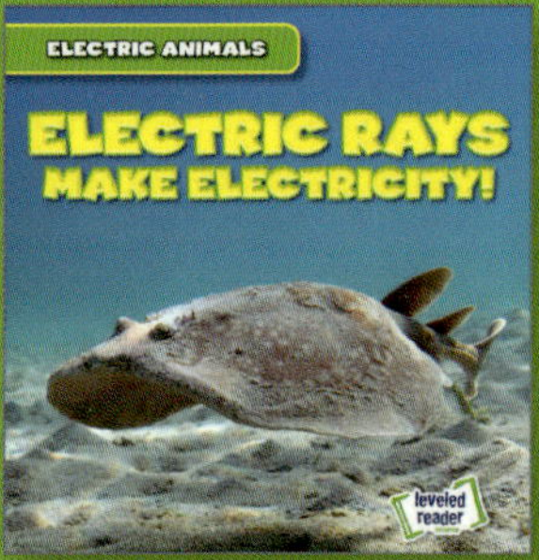

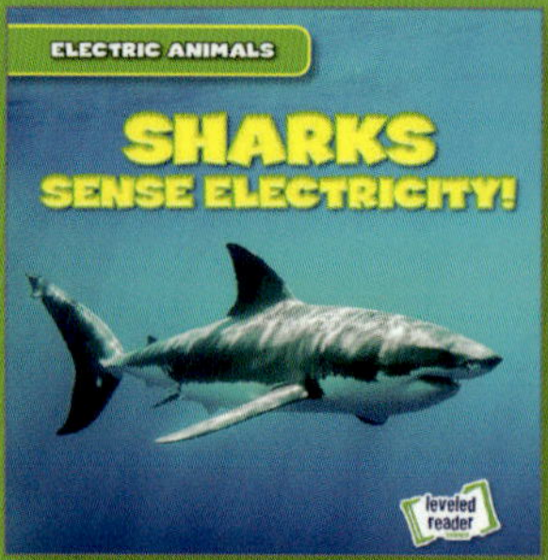

ISBN: 9781538292938

9 781538 292938

Gareth Stevens PUBLISHING